D1174823

Reading Essentials in Science

LIVING WONDERS

ADAPTATION and SURVIVAL

SUSAN GLASS

PERFECTION LEARNING®

Editorial Director: Susan C. Thies
Editor: Mary L. Bush
Design Director: Randy Messer
Book Design: Michelle Glass
Cover Design: Michael A. Aspengren

A special thanks to the following for his scientific review of the book:

Paul Pistek, Instructor of Biological Sciences, North Iowa Area Community College

Image Credits:

©Bettmann/CORBIS: p. 8; ©Galen Rowell/CORBIS: p.10 (top); ©Lawson Wood/CORBIS: p. 16; ©Stephen Frink/CORBIS: p. 19; ©Robert Pickett/CORBIS: p. 21 (middle); ©Neil Miller; Papilio/CORBIS: p. 31 (bottom); ©Kevin Schafer/CORBIS: p. 32 (top); ©John Holmes; Frank Lane Picture Agency/CORBIS: p. 33

Photos.com: front cover, back cover, background on all pages, sidebar background, pp. 2, 3 (far left & far right), 4, 5 (top), 6 (left), 7, 10 (bottom right), 12, 13, 14, 15 (top & bottom), 17, 20, 21 (top & bottom), 23, 24, 25, 26 (top), 27 (bottom left & right), 28 (top left), 29, 30, 31 (top), 32 (bottom three), 34, 35, 36, 37, 38, 40, 41, 42, 43 (middle), 45 (right); Corel Professional Photos: pp. 10 (bottom left), 11 (bottom), 18 (top), 26 (bottom), 28 (top right), 39; Ingram Publishing: p. 3 (middle); Photodisc: pp. 6 (right), 43 (far left & far right); Digital Stock: pp. 18 (bottom), 22, 45 (left), 48; Eyewire: p. 27 (top); Hemera Studio: p. 9; Perfection Learning Corporation: p. 28 (bottom); Tobi Cunningham: p. 11 (top); Emily Greazel: p. 15 (middle)

For information, contact
Perfection Learning® Corporation
1000 North Second Avenue, P.O. Box 500
Logan, Iowa 51546-0500.
Phone: 1-800-831-4190
Fax: 1-800-543-2745
RLB ISBN-13: 978-0-7569-4479-7
RLB ISBN-10: 0-7569-4479-1
Paperback ISBN-13: 978-0-7891-6301-1
Paperback ISBN-10: 0-7891-6301-2
4 5 6 7 8 PP 15 14 13 12 11
perfectionlearning.com
Printed in the U.S.A.
PP / 03 / 11

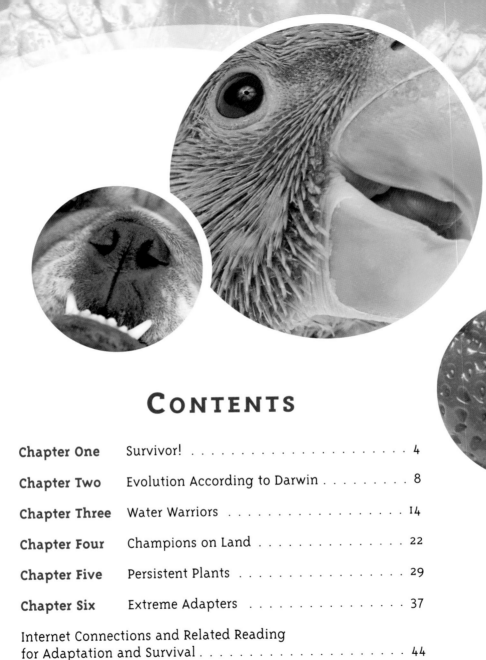

CONTENTS

chapter one

Survivor!

One in a Million

The word *species* ends in *s* whether it is singular or plural. You are a member of one species, but there are millions of species in the world.

THE STORY ON SPECIES

● ● ●

WHAT MAKES YOU different from a dog or a frog or a pine tree? Each of you is a member of a different **species**. What is a species? A species is a particular type of living thing. Members of a species resemble one another and can **mate** to produce more members of the same species. Humans are a species and so are cats, mosquitoes, and dandelions (and dogs, frogs, and pine trees).

Even though you and your parents and your best friends are all members of the same species, does that mean you're exactly alike? Of course not. Members of a species share certain common characteristics, but there are also differences, or **variations**, among them. For example, all humans walk on two feet, use two hands, and have a brain and a heart that function the same way. However, some people are tall and some are short. Some are thin and some are heavy. Some have red hair and some have black hair. Some have brown eyes and some have blue eyes. These variations make each member of the human species unique.

Over time, changes in a population's characteristics will occur. Birds may develop different-shaped beaks. The color of a bear's fur may change. Fish may produce their own light to see in dark waters. Species are constantly changing, but these changes don't happen overnight. They happen over long periods of time, one generation at a time.

Try This!

Dump out a pile of peanuts in their shells. Ask a few friends to take a couple of peanuts and draw pictures of them. Put all the peanuts back in the pile. Can you identify each peanut from its drawing? Even though all of the peanuts have a crunchy outer shell, are a brownish color, and taste alike, there are still differences in each one. No two are exactly the same.

THE GAME OF SURVIVAL

• • •

Why do species change? To survive! Survival is a challenge. It is a serious game with winners and losers. Those species that win continue to live on Earth. Those that don't, die off.

Why do some species live and some die? The world is always changing. Mother Nature throws a lot of curves. She might send a drought or an ice age or a flood or an epidemic. Other times, new species cause the disappearance of existing species. A new **predator** with big teeth or new competitors that gobble up the food supply can be dangerous to weaker species. Humans are also responsible for some of the changes on the Earth. They cut down trees in the rain forest, pollute river water, and clear out **habitats** to make room for malls and businesses. It's a tough world out there, and if a species wants to survive, it has to **adapt**.

To adapt means to change and adjust to new circumstances or conditions. All living things are constantly adapting to their **environments**, or surroundings. Over time, generations of species adapt in order to survive when the environment changes. Those who can't successfully adapt don't make it. They become **extinct**.

Adaptations are characteristics that help an **organism** survive in its environment. Adaptations can be physical properties (how something looks) or behaviors (how something acts).

Adaptations are inherited, or passed down from one generation to the next. A frog can't just change its color from green to brown one day because it decides to. But if a common green frog mates with a rare brown frog to produce a new brown frog and that brown frog mates with another brown frog and so on and so on, over time, the species may change from green to brown.

Sometimes these variations within a species improve an organism's chances of survival. For example, imagine that the forest where the green frogs live experiences a climate change. It becomes hotter and drier in the forest. The green trees wither and die. Now the bright green frogs are easily spotted and eaten by predators. The few brown frogs are able to remain hidden in the dry brown leaves, branches, and tree trunks. Eventually, the green frogs will die out, and the brown frogs will survive and **reproduce**, filling the forest with more brown frogs.

Adaptations are the key to an organism's survival. Those that are able to adapt to their changing environment will survive. They will outwit and outlast those organisms that can't adapt.

chapter two

Evolution According to Darwin

A CHANGE IN A species over time is known as **evolution**. All living things on Earth today are evolved, or changed, forms of living things that came before them.

Many scientists have had ideas about how evolution works, but the man whose theory is most widely accepted today was an English scientist named Charles Darwin. Darwin presented his theory of evolution to the world in 1858 after he'd spent more than 20 years gathering evidence and putting the pieces together.

SPECIES AROUND THE WORLD

• • •

Darwin began laying the foundation for his theory of evolution in 1831 when he set sail on a British naval ship named the Beagle. The young Darwin became the ship's naturalist. It was his job to study nature at every stop the ship made. Darwin was awed by the great variety in the plants and animals he saw along the journey.

The Beagle sailed around the world, stopping in places like New Zealand, Australia, and South America. Darwin saw many plants and animals that he had never seen before. He wondered why they were so different from those he'd seen in Europe.

In South America, Darwin observed hairy sloths hanging upside-down in trees. He also collected fossils of ancient sloths. When he compared the living sloths to the fossils, he noticed that some of the fossilized bones were much larger than those of the living sloths. He wondered why the size of sloths had changed over the years.

If He'd Only Known . . .

Darwin would have been a lot more awed if he'd known just how much variety exists among living things. Scientists have now identified over 2.5 million species of organisms living on Earth today.

Iguana

Then in 1835, the Beagle sailed into the Galápagos Islands. These small, isolated islands are hundreds of miles off the west coast of South America. Darwin was fascinated with the plants and animals he found there. While many were similar to those that he had observed on land in South America, he also noted variations. For example, there were iguana lizards on the islands that looked a lot like ones on the continent. But the island ones swam in the ocean and ate seaweed. The ones on the mainland climbed trees and ate leaves. Both had body differences suited to their lifestyles. There were also large sea birds called *cormorants* in both places, but the ones on the continent could fly while the ones on the islands could not.

Tortoise Islands

The Galápagos Islands are named for the giant tortoises that live there. *Galápagos* means "tortoise" in Spanish.

Cormorant

Tortoise

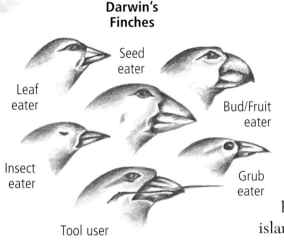

Darwin's Finches

Leaf eater

Seed eater

Insect eater

Bud/Fruit eater

Grub eater

Tool user

Darwin also realized that the plants and animals differed from one island to another. He observed small birds called *finches* on several islands. Finches on one island ate seeds and had strong, wide beaks. Finches on another island had long, pointy beaks for eating insects. Darwin recognized that each finch had a beak that matched the type of food it ate.

DARWIN'S NATURAL SELECTION
● ● ●

After he returned to England, Darwin organized his information about plants and animals, gathered more facts about changes in the Earth, and pieced together his theory of evolution. He finally concluded that the species on the Galápagos Islands had changed over many generations to become better suited to their environment. The birds on the mainland in South America only had one type of diet, so they only needed one kind of beak. When the birds drifted to the Galápagos Islands, they encountered new environments and new food sources. In order to survive, the birds' beak shape gradually changed to fit their diet. Darwin believed this was true of other plants and animals as well.

Galápagos Islands

Darwin called his theory of evolution *natural selection*. Natural selection is a blend of several ideas.

- Most species produce more **offspring** than can possibly survive. Since food, water, and other resources are limited, the offspring must compete with one another to get what they need.

- There is variation within a species. Each organism has different characteristics. Some of these characteristics make certain individuals better suited to the environment. These members will win the competition for resources.

- The "winning" organisms will live long enough to reproduce more offspring with the same characteristics. In other words, those individuals with characteristics that help them survive are selected by nature to pass on those characteristics to future generations.

Natural Selection in Action

Imagine that a company clears part of a forest to build a vacation resort. Most of the short trees are removed. All that remains are the tall trees on the edges of the forest. A herd of giraffes lives in the forest and depends on the trees for food. Now only the giraffes with the longest necks will be able to reach the leaves. The short-necked giraffes will starve and die. The long-necked giraffes are more likely to survive and reproduce. In following generations, more of the herd will have longer necks.

SURVIVAL OF THE FITTEST

● ● ●

Sometimes Darwin's ideas are summed up in the phrase "survival of the fittest." You probably take P.E. class to "get fit." You run laps, do sit-ups, and play sports. So does that mean that animals that can do pull-ups and outrun predators in the 100-yard dash will be the survivors? Not necessarily. The "fittest" in Darwin's theory are those best suited for or best adapted to the conditions in the environment at the time. They might be the moths with coloration that helps them blend in with their background so birds don't see and eat them. They may be the plants that have thorns so animals take one bite and walk away.

Darwin's fittest aren't necessarily the biggest, fastest, or strongest. Sometimes they are even the smallest or the slowest. Small walking sticks hide from danger by looking like the branches they perch on. South American tree sloths move so slowly that predators don't notice them. While perhaps not the most physically impressive, these organisms will survive because they fit best in their given environment.

THE ULTIMATE SURVIVOR

● ● ●

Over millions of years, living things have adapted to the highest mountains, the deepest oceans, the driest deserts, and the coldest ice caps. Every species alive today is the result of adaptation. These organisms have found ways to get the resources they need from their environment and to keep from dying long enough to reproduce. They have become the ultimate survivors.

chapter three

Water Warriors

A QUICK GLANCE AT A GLOBE or world map will remind you that most of the Earth is covered with water. Organisms that live in or near water need special adaptations for survival.

IN THE OCEAN
• • •

The oceans are full of organisms that have adapted to life in salt water. Oceans have several levels, or zones. The conditions are different in each zone, so organisms in each zone have different adaptations.

The top zone, or sunlit zone, is located at the surface of the water. Because sunlight reaches this level, plants and other organisms can use the Sun's energy to produce food. **Phytoplankton** and **algae** thrive in the sunlit zone. Many fish, shrimp, and whales feast on these organisms.

Beneath the sunlit zone is the twilight zone. This area of the ocean receives only a little sunlight, so plants and other organisms that need sunlight to make food cannot live here. This reduces the food supply for animals in this zone. The animals must also adapt to life with little or no light. In addition, they must be able to survive the colder temperatures and higher water pressure. Many fish in this zone have special adaptations such as sharp teeth and large mouths and eyes to help them find and capture food in the dark.

Sunlit zone

Twilight zone

Midnight zone

No light reaches the bottom of the ocean, known as the midnight zone. The animals that live here must overcome tremendous water pressure, near-freezing temperatures, and little food. Many of these creatures use smell, sound, or vibrations in the water to find what little food exists.

Fish Features

Fish have many adaptations for underwater survival. They have sleek, streamlined shapes for slipping through water. They have fins for smooth swimming motion. A slimy coating of mucus also helps fish slide through the water. A covering of scales provides protection and waterproofing.

Fish breathe with gills that take oxygen out of the water. They have nostrils that are adapted for smelling. Eyes on the sides of their heads enable them to see what is in front, to the sides, and behind them. They even have lines of sensory pores called *lateral lines* on their sides to help them detect movement.

Flatfish

Flatfish spend a lot of time lying on their sides on the seafloor. Both of their eyes are located on the side of the fish that faces upward. This adaptation helps them spot trouble coming from above.

Most fish have an air sac inside called a *swim bladder*. This sac fills with gases so the fish can rise in the water. When the fish wants to sink to lower depths, the gases empty from the bladder.

Fish also use **camouflage**, or protective coloring, as a means of adaptation. They have light-colored bellies and darker backs. The light color camouflages, or hides, them in the light when a predator looks up at them. The dark upper part of their bodies camouflages them against the dark ocean depths when a predator looks down at them from above.

The plaice is a type of flatfish with bright spots on its upper side and a white underside.

Pufferfish

About 120 species of pufferfish live in tropical oceans around the world. These fish get their name from their special adaptation. When they are scared, pufferfish gulp in air and water and blow themselves up like a balloon. All puffed up, they're too big for predators to eat. One pufferfish, the porcupine fish, is covered with spines that stick out when it inflates itself. Others have poisonous inner body organs that can kill a person who eats them.

In Japan, raw pufferfish is considered a gourmet treat. Chefs carefully cut out the poisonous organs and slice the flesh into thin pieces. Tragically, even with careful preparation, eating pufferfish still kills a few Japanese diners each year.

The Underwater World of Whales and Dolphins

Whales and dolphins look like fish but are actually **mammals**. They have fins and streamlined bodies for smooth swimming motion. Instead of noses, they have blowholes on top of their heads that close underwater.

Whales have thick layers of fat called *blubber* that keep them warm in cold ocean waters. Toothed whales have teeth for catching and eating fish, squid, and other ocean animals. Baleen whales have bony fringes instead of teeth that strain tiny **prey** out of huge mouthfuls of water.

Squiggly Squid

Like fish, some squid have darker upper bodies and lighter undersides for camouflage. Squid can also send waves of color down their bodies to disguise their outline from enemies. If a predator still comes too close, squid squirt a cloud of dark ink to hide behind as they make a quick getaway.

Shocking Eels

Octopus

Electric eels can only eat prey that isn't moving because they have no teeth to trap food with like most predators. So what *do* they do? These creatures send out electric shocks. Special muscles in their tails produce enough electricity to kill a small fish and stun larger fish and humans.

Eight Is Better Than Two

The octopus has a whole bag of tricks for eating and keeping itself from being eaten. This animal has eight arms with rows of suckers for grabbing and holding on to prey. Octopuses hide from predators by changing their color to blend in to their surroundings and by squirting an ink cloud to hide behind.

Octopuses also have a poisonous bite. The most dangerous octopus is the blue-ringed octopus from Australia. While it is only six inches long, its bite can kill a human within five minutes. The blue-ringed octopus does try to warn predators though. If it feels threatened, it flashes its bright blue rings to scare off a predator before it chomps into it.

Fierce Fighter

The Portuguese man-of-war is not one animal, but hundreds of tiny organisms living and working together. Each tiny organism has a special job. Some help with floating, others digest food, and others sting prey. The strange and frightening Portuguese man-of-war has a large float on top with 50-foot-long tentacles hanging from it. The float can change shape to make the most of wind movement. The tentacles are loaded with stinging cells that poison prey.

Portuguese man-of-war

Getting Even

Even the fearsome man-of-war isn't safe. There is a sea slug that has adapted itself to be **immune** to the man-of-war's poison. This slug can eat the man-of-war and steal its stingers to use as weapons for itself.

IN FRESHWATER

Freshwater plants and animals live in or around rivers, streams, lakes, and ponds. Most plants and animals in these bodies of water live in the shallow water where sunlight penetrates. This enables the plants to use **photosynthesis** to produce food for themselves. Many animals can then feast on these growing plants.

Dragonfly

Water strider

Many of the birds, insects, and animals that live near freshwater have adapted to life both in and out of the water. Ducks, for example, have strong webbed feet for paddling through water, but they can also waddle along the shore. Dragonflies begin life in the water, where they remain until they are fully grown and can fly to escape predators. Water snakes are excellent swimmers but also slither along the ground and in the branches of trees on the water's edge. Frogs have lungs to breathe on land but can also take in oxygen through their moist skin and the lining of their mouths when in the water. Alligators and crocodiles have webbed feet and powerful tails to propel themselves through water, but they can also be found crawling along the banks of rivers and swamps. These dual adaptations provide these animals with a wider variety of survival options.

Look Out, Olympic Skaters!

Water striders are also known as pond skaters because of their amazing ability to glide across the water.

Water Striders

Water striders are insects with several special adaptations for water. These insects are able to run across the surface of ponds and streams. Their long middle legs push them across the water surface while their long hind legs steer and brake. Two short front legs snatch prey. Water striders can also sense motion on the water's surface, which helps them locate prey.

Beavers

Beavers have many adaptations for living on land and in water. Strong jaws and sharp front teeth enable them to eat the bark off trees on the riverbank and gnaw on wood to make dens (houses) in the water. A paddle-shaped tail provides support and balance on land and makes them skilled swimmers in the water. When these animals dive underwater in search of plants on the bottom of a river, they have valves that close off their ears and noses and a flap of skin that seals off their mouths (except for their teeth).

Beaver

Crayfish

Crayfish

What happens when a crayfish loses a leg? It grows a new one! That's quite a handy adaptation to have when you spend your life sticking your legs into cracks in rocks looking for food. Crayfish also have strong pinchers for getting food and defending themselves. The hard outer skeleton of a crayfish is another protective adaptation.

Cattails

Cattails are found in shallow ponds, freshwater marshes, and on the edges of calm lakes and rivers. These plants grow and spread quickly, increasing their population easily. They are able to do this because they have special leaf and stem cells that draw in large amounts of oxygen. This is important because often the oxygen supply for plants is low in areas of slow-moving water.

Cattails

chapter four

Champions on Land

LIVING THINGS ON LAND HAVE THE SAME BASIC NEEDS AS THOSE living in water. Their different environment, however, means that they need different adaptations for getting food and protecting themselves.

FEED ME!
● ● ●

Before an animal can eat, it has to locate and catch its food. An eagle's eyesight is three to four times stronger than a human's eyesight. Eagles can spot prey from great distances and then swoop down and grab it with their large talons (claws). Owls can see in the dark to help them hunt at night. They can also fly without making a sound so they can sneak up on prey. Bats find food in the dark using echolocation. They bounce sounds off prey to locate it. Camouflage also helps predators sneak up on prey. White polar bears blend into their snow and ice environments. They can sneak up on seals and make a meal out of them.

Bald eagle

Whatever the food of choice is, there is a mouth adapted for eating it. Plant-eating mammals have flat teeth adapted for grinding up plants. Meat eaters have pointed teeth for tearing meat. Those that eat both plants and animals have a mixture of flat and pointy teeth. Frogs and anteaters have long, sticky tongues for catching insects.

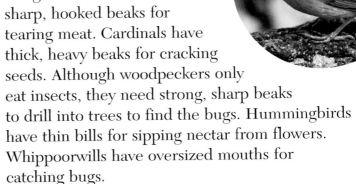
Cardinal

Bird beaks come in many shapes and sizes, all fitted to the birds' diets. Eagles and hawks have sharp, hooked beaks for tearing meat. Cardinals have thick, heavy beaks for cracking seeds. Although woodpeckers only eat insects, they need strong, sharp beaks to drill into trees to find the bugs. Hummingbirds have thin bills for sipping nectar from flowers. Whippoorwills have oversized mouths for catching bugs.

Woodpecker

Hummingbird

Finding Your Type

Animals that eat plants are called *herbivores*. Animals that eat other animals (meat) are *carnivores*. Animals that eat both plants and meat are called *omnivores*. Which one are you?

I Want to Suck Your Blood!

Ever wonder where that expression came from? Vampire bats have special adaptations for getting their dinners. They suck blood from pigs, horses, and cows for nourishment. Their teeth are so tiny and sharp that their bite doesn't hurt and their victims sleep right through the attack. When the blood starts to flow, the bats lap it up with their tongues. Bat saliva has chemicals in it that mix with the blood at the wound site and keep the blood from clotting. That way the bats can drink all they want.

STAY AWAY!

• • •

Body structures, behaviors, camouflage, and **mimicry** are types of adaptations that help animals protect themselves.

Back-Off Body Adaptations

Many animals have well-developed senses of smell, sight, and hearing to alert them to danger. Large ears, eyes, and noses are often clues that an animal has extra sensing abilities. Deer, kangaroos, and rabbits have long legs to help them outrun or outhop their enemies. Stingers, spines, claws, tusks, horns, and shells are also protective adaptations. Very few predators want to approach a prickly porcupine or make a rhinoceros with a giant horn angry!

Smelling or tasting bad also serves to protect. Stinkbugs didn't get their name for nothing! These bugs are named for the odor they give off when threatened. Skunks spray a stinky substance into the air when they are frightened or angry. The American toad produces a chemical that makes it taste yucky to predators. Monarch butterflies taste bitter. Once a bird has tasted one, it usually won't try another.

I Warned You

Some animals with bad taste or smell don't need camouflage. Instead they often advertise themselves in bright colors to warn predators to stay away. Skunks are clearly visible in black and white. Monarch butterflies are a bright orange color.

Poison is another defense adaptation. There are poisonous snakes, poisonous fish, poisonous caterpillars, and poisonous birds. A shrew is a mammal with a poisonous bite. The Mexican beaded lizard and its cousin the gila monster are poisonous lizards. Wasps and bees have poisonous stingers. Some beetles shoot poisonous chemicals at predators. Amphibians (frogs, toads, and salamanders) don't like to fight, so they have special poison-making glands in their skin. Usually the poison just makes these creatures taste bad, but a few species are dangerous. Poison-arrow frogs of South America have a poison so deadly that Native Americans coat the tips of blow darts and arrows with it.

Newts are members of the salamander family.

The Best Behaviors for Survival

How an animal acts can save its life. Animals that live together in groups, such as flocks of birds or herds of sheep, protect one another. Other animals, such as wolves, hunt together in packs. Musk oxen form a circle when threatened by wolves. The young members of the herd hide inside while the adults use their horns and hooves to warn attackers to stay away. Swallows, starlings, and other songbirds swarm around hawks and other predators to harass them into leaving the group alone.

Some animals have individual behaviors that protect them from harm. Rabbits freeze when they think they've been spotted. Grass snakes and opossums play dead to fool predators. The hawk moth caterpillar waves its body like a snake, scaring away birds.

When an environment changes temporarily and an animal can't adapt, it can migrate. **Migration** is the movement of an animal to a new habitat for a period of time. Animals migrate to escape harsh weather, to find food, and to reproduce. Geese and other birds migrate south to warmer temperatures in the winter. Monarchs migrate to Mexico from Canada and the United States. Caribou, elk, and elephants migrate to find food during cold or dry seasons. Termites, Japanese beetles, and earthworms migrate down farther in the soil in colder temperatures. Umbrellabirds migrate to the tops of tall trees to reproduce and then move down to lower ground the rest of the year.

Japanese beetle

Are We There Yet?

The Artic tern travels over 20,000 miles each year to migrate. These birds spend the summer in the Artic and then fly south to spend winters in the Antarctic.

Arctic tern

Catch-Me-If-You-Can Camouflage

Many animals use camouflage for protection. Predators use it to sneak up on prey, and prey use it to hide from predators. The stripes on zebras are designed to confuse lions, their biggest predators. The wavy black and white stripes of the zebra blend in with the wavy grass around them. A tiger's stripes help it hide in the shadows of the rain forest. The Arctic fox changes from brown to white when the snowy season begins.

Lions are colorblind, which means they don't see things in color. Everything is black and white. That's why even though a zebra's stripes are black and white and grass is brown or green, a lion can't spot a zebra. It's the wavy pattern on the zebra, not the color, that camouflages it.

Chameleons, octopuses, and squid all change color to match their surroundings. The two-toed sloth has long hair with grooves in it where algae can grow. The algae turns the sloth's coat a greenish color that is perfect for blending in with rain forest trees.

Some insects disguise themselves by looking like an object in nature. A walking stick looks like a twig. Grasshoppers, praying mantises, and cryptic frogs look like the leaves they rest on. Many tree frogs have skin that looks like tree bark.

Praying mantis

Viceroy

Monarch

Disguised for Safety

Mimicry is an adaptation for disguising oneself by looking like another species that is harmful or not of interest to predators. Flies that look like bees are ignored by birds who don't want to get stung. The viceroy butterfly gets left alone because it looks like a poisonous monarch butterfly. The harmless scarlet king snake is avoided by predators because it looks like a deadly coral snake. Hummingbird moths look like hummingbirds so they aren't eaten by other birds. Some caterpillars, moths, and butterflies have spots on them that look like big eyes. This makes them resemble threatening predators.

Try This!

Camouflage a potato. Use a small potato, stick pins or glue, and a variety of materials, such as dried beans, seeds, feathers, leaves, and straw.

Choose an environment for your potato "animal." It might be outside in a tree or in the grass or indoors on a rug or on your bed. Decorate your potato with materials that will help it blend in to its environment. For example, if you want it to blend in with your blue bedspread, then you might cover it with blue feathers, beads, or buttons. Share your camouflaged creature with your class, and show or describe the environment it's adapted to.

chapter five

Persistent Plants

YOU MAY THINK OF PLANTS AS HARMLESS, green leafy things that decorate your house or yard, but these amazing living things have numerous adaptations that enable them to survive in this tough world.

ROOTS, STEMS, AND LEAVES

• • •

Location and climate affect different root adaptations. In forests, plants have deep roots to anchor them in the ground. They may even have stilt or prop roots above the ground to help keep them from blowing over in strong winds. Some plants have one thick root that grows straight down called a *taproot*. Taproots reach deep into the ground for water and sometimes store food as well. If you've ever pulled a dandelion weed, you might have noticed its long taproot. In tropical rain forests, some plants grow on branches high in the trees. Their roots hang on to the tree branch and absorb water from the moist air.

Turnips

Sugarcane

Some plants have adapted to weather conditions by storing food and water. When weather conditions are right, they make extra food and take in extra water. Some plants store the extra supplies in their roots, and others store it in their stems. Sugarcane is sweet because of the stored food (sugar) in its stems. Sugar beets are roots that hold stored sugar for the beet plants. When you eat sweet potatoes, turnips, beets, carrots, or potatoes, you are eating the food a plant was saving in its roots for itself.

The leaves of plants are also the result of adaptation. Some are huge. Some are tiny. Some are rounded. Some are jagged or pointed. Leaves can be found in hundreds of different shapes and sizes, each suited perfectly to its environment.

Most leaves are thin and flat to absorb sunlight and water so photosynthesis can take place. Most trees shed their leaves in the fall to help them survive a winter with little food and water. Evergreen trees, on the other hand, have needle-shaped leaves that reduce water loss and help them survive the winter months. The spines on a cactus are actually modified leaves that discourage animals from eating the plant. On some cactuses, these spines may also help it endure dry desert conditions by collecting morning dew.

Plant Carnivores

Some plants can't get the nutrients they need from the land, so they have adapted to eat meat (bugs). The Venus flytrap has leaves that snap shut when an insect lands on them. Then the plant digests the bug to get the nutrients it needs. The pitcher plant has pitcher-shaped leaves filled with liquid. The plant has a sweet smell that attracts bugs. When they land on the plant, they slide down the slippery insides of the "pitcher," drown in the pool of liquid, and get digested by the plant.

Venus flytrap

Strangler fig

Tendrils are adapted leaves or stems found on certain plants. These threadlike parts of climbing plants support the plant by clinging to or coiling around an object such as a tree trunk or branch. Clematis plants use their leaves to twist around twigs or stems of other plants. In areas where competition for sunlight is fierce, climbing plants use their tendrils to help them climb above other plants to reach the Sun's rays. Lianas and strangler fig trees are examples of rain forest climbing plants.

Built-In Firefighters

Many prairie plants have adapted to fire. Prairie grasses have growing structures below the ground that are protected from fires. Many oak trees have very thick bark that doesn't burn easily. Other trees and shrubs have underground parts that quickly resprout after a fire.

Oak

PARASITES

• • •

Some plants survive by stealing food from other plants. A parasite is an organism that lives off another organism. Mistletoe, dodder, and rafflesia are examples of parasitic plants. Mistletoe lives in trees and steals their water, food, and minerals. Dodders are parasitic weeds that attach themselves to other plants and spread quickly. Rafflesias are giant plants that live off other plants on the rain forest floor.

The World's Largest Parasite

The rafflesia plant grows the largest flowers in the world. These flowers can weigh almost 40 pounds and measure almost 40 inches across.

Rafflesia

PROTECTION

• • •

Plants are food for many animals, so they are in constant danger of being eaten. Plants can't run away, so they have come up with other ways to protect themselves.

Sharp structures on plants keep animals away. Roses have thorns. Thistles have spines. A holly bush has leaves with spikes. The holly leaves that are closer to the ground are even spikier than those at the top of the bush since more animals can reach the lower leaves.

Plants can also protect themselves from hungry animals by tasting bad or being poisonous. Redwood trees, for example, have bad-tasting bark that insects avoid. Ivies, bleeding hearts, and lillies of the valley might be pretty to look at, but they are poisonous when eaten. Some trees produce poison in their leaves only when they are being munched on. The leaves of Indian millet, also called *sorghum*, contain a poison that only becomes active if an animal takes a bite of a leaf.

Holly

Redwood

Rosebush

REPRODUCTION

• • •

Since plants can't move around the way animals do, some have adaptations to get animals to spread **pollen** so the plants can reproduce. Plants are masters at tricking animals. The colors, fragrances, and sweet nectars of flowers attract insects and other animals. The animals travel from flower to flower, unknowingly collecting and dropping off pollen as they go. The rafflesia flower smells like rotting dead animals. Why? Its stink attracts hungry flies that then spread pollen for the plant. Some orchids have flowers that look and smell like female bees. The flowers attract male bees that are looking for a mate. The males don't get a mate, but they *do* pick up pollen and accidentally carry it to other flowers.

Coconut palm

Once plants have produced seeds, it's important that they get scattered away from the parent plant. Otherwise, the new plants are overcrowded and their chances of survival drop. So plants have come up with brilliant ways to disperse, or spread, their seeds. Dandelions, milkweeds, and other plants produce seeds with fluff that helps them float in the breeze. Some plants, such as violets and Himalayan impatiens, shoot their seeds into the air like little cannonballs. Other seeds, like those of maple and pine trees, have wing-shaped parts that help them glide to new homes. Palm trees drop coconuts into coastal waters. These huge seeds are adapted for floating to new habitats. When a coconut gets washed ashore on a sandy beach, it can sprout and grow into a new palm tree. Did you ever see tumbleweeds rolling across the land in a Western movie? That tumbling action is another adaptation for dispersing seeds.

Sometimes plants use animals to disperse their seeds. Raspberry and strawberry seeds are attached to fruit that birds eat. The seeds are carried away from the parent plant and left behind in the birds' droppings. Oak trees use squirrels to help them scatter their seeds (acorns). Squirrels carry off acorns and bury them in hiding places to eat later. Some of them sprout in their new location before the squirrels return to eat them. Have you ever had a burr stuck to your socks? You were being used by its parent plant. Those burrs are adaptations for using animal fur (or human clothing) to carry seeds away from the parent plant.

chapter six

Extreme Adapters

Extreme environments can challenge even the best adapters. Still, many plants and animals have managed to adapt to the coldest, darkest, wettest, driest, and hottest places on the planet.

BRRR!
IT'S COLD
● ● ●

Some animals that live in extremely cold temperatures use an adaptation called **hibernation**. Because food is scarce and temperatures are freezing, bears, skunks, chipmunks, snakes, bats, and other animals can go into a deep sleep. They require very little energy in this hibernating state and are able to live off stored fat until warmer weather arrives.

Plants in cold habitats have also adapted to the temperature, wind, snow, and ice. Some of these plants have deep roots that anchor them in the near-freezing ground. Some are very flexible so they can bend without breaking in strong winds. Limber pine and white bark pine trees are two "bendable" trees found in **alpine** environments. Many alpine trees grow in short, stunted, shrublike versions called *kruppelholz* to keep them out of the icy wind. Having a slick or waxy covering keeps some plants from being buried under mounds of snow. For example, snowflakes slide right across the shiny leaves of the mountain alder.

Antarctica's chilly waters support a surprising amount of plant and animal life. Phytoplankton support a dense population of tiny shrimplike krill, which in turn feed many larger animals like penguins, whales, and seals. A hundred feet down in the water, food is scarce and temperatures are frigid. Animals have adapted by slowing their bodies down so they don't require as much food and heat energy. Sponges can live for many centuries in the Antarctic waters. The ice fish, a bottom dweller, has adapted to the cold by producing up to eight kinds of antifreeze substances to keep its body fluids flowing in subfreezing temperatures. Seals are covered with thick fur and a layer of blubber (fat) that keep body heat from escaping.

Polar bears hunt for seals in the dazzling whiteness of the Arctic. The bears keep from getting "snow blindness" by closing the pupils of their eyes to the size of pinholes. Polar bears also have a heightened sense of smell. They can detect a seal several miles away. The skin of polar bears is black to help soak up the heat of the Sun. Their fur is made of hair that is hollow in the middle to better insulate the bear and keep it warm.

Frost-Free Fur

The fur of wolverines is frost-proof. Not only does it keep these animals warm and dry during harsh temperatures, but it is also used to make fur trim for parka hoods because it prevents a person's breath from changing to water on the fur.

IT'S DARK DOWN HERE
● ● ●

Beneath the open ocean is a cold, dark world. Here creatures often have large eyes to see in the darkness. Giant squid have the largest eyes of any animal on Earth—as large as ten inches across!

Some creatures in this dim habitat glow in the dark. Bioluminescence is light produced by living things. Glowing in the dark helps animals find food and mates. Some animals also use this light to frighten predators or even to camouflage themselves. Black dragonfish, jellyfish, **copepods**, krill, and some types of **plankton** and **bacteria** are bioluminescent.

Most fish swimming near the ocean bottom are predators with big mouths and oversized stomachs. Some swim around with their mouths wide-open all the time, hoping to scoop up any available food. Anglerfish and some eels have huge stomachs that can hold prey larger than the fish themselves. This helps them fill up when they can.

The deep-sea anglerfish is a super adapter. This fish has an enormous mouth lined with long, curved teeth. These teeth fold backward into the mouth to let prey in and then spring back upright to keep the prey from escaping. Anglerfish also have a built-in fishing rod and bait to lure prey in the dark. The rod is a spine attached to the fish's head. The bait is a piece of skin dangling from the end of the spine. Millions of luminous bacteria live on the "bait" and make it glow in the dark with a bluish or greenish yellow light. When it gets hungry, the fish waves the bait back and forth and attracts fish, which it then gobbles up quickly.

Fishing for a Mate

The word *angler* means "fisherman." Since mates are hard to find on the dark ocean floor, the male anglerfish attaches itself to any female he manages to find. Their bodies grow together, and the male fish is supported by the female.

I'M HOT AND THIRSTY!
● ● ●

Plants and animals in the desert have adapted to the heat and lack of water. Many desert plants roll up their leaves during the hottest part of the day or grow in locations where the Sun does not hit them directly. Some plants have thick, waxy skins to keep water from escaping. Long taproots help some desert plants seek water deep in the ground.

Cactus plants have shallow roots that spread out far from the plant, ready to soak up the rain when it finally falls. Cactuses store water in their thick stems.

Avoiding heat is a top priority for desert animals. Many creatures, such as foxes, skunks, and bats, are nocturnal—they come out at night when it's cooler and sleep in the shade during the day. Desert amphibians and **rodents** burrow into the ground to escape the Sun.

Fox

A few animals estivate, or sleep, during the hottest time of the year. Estivating is like hibernating, only in hot weather.

The animals in this hot habitat also have adaptations for finding and saving food and water. The gila monster lizard stores fat in its tail and lives off it when food is scarce. Camels also store food in the form of fat in their humps. They can also go long periods of time without water. The main sources of water for most desert animals are the plants they eat. However, some animals have other sources. The thorny devil lizard is covered from head to foot with spikes. When dew forms on these spikes, it trickles down to the lizard's mouth. Kangaroo rats seal off their dens and reuse moisture from their own breathing.

THE ULTIMATE ADAPTERS
● ● ●

Humans are the ultimate adapters. When temperatures grow cold, they build shelters. When it gets hot, they turn on the air-conditioning. In times of drought, they use irrigation. In times of floods, they build barriers. With intelligence, skill, and communication, the human population has managed to multiply and spread out over the globe.

However, humans' success in meeting their needs and wants has changed the environment for other species. Humans have chopped down forests for farmland. They've cleared rain forests to build roads and cities. They fish the oceans, hunt animals on land, and pollute the air and water. Humans change and destroy habitats every day.

Human actions have made survival difficult or impossible for many species. Some plants and animals have not been able to adapt to their rapidly changing or disappearing environments. Some have become endangered, and others have become extinct. And while extinction is part of Darwin's natural selection process, the recent increase in the extinction of species is believed to be due to human behavior, not nature.

It is important that humans understand the effect they have on the survival of other species. Guaranteeing a wonderfully diverse world for the future begins with choices made today.

Oil spills injure or kill ocean life.

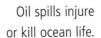
Logging destroys forest habitats.

Spraying pesticides on crops pollutes groundwater that is used by humans and animals.

INTERNET CONNECTIONS
AND RELATED READING FOR
ADAPTATION AND SURVIVAL
● ● ●

Internet sites

http://www.cotf.edu/ete/modules/msese/earthsysflr/adapt.html
Visit the "Earth Floor" and learn how plants, animals, and humans adapt to the changing environment.

http://ecokids.earthday.ca/pub/eco_info/topics/climate/adaptations/index.cfm
Be an "EcoKid." Play the animal adaptation game, take the Arctic adaptation challenge, and learn more about animal antics.

http://www.nisd.net/leonsww/learninglinks/oceans/oceans.htm#twilight
Explore some unique adaptive creatures found in each of the ocean zones.

http://www.desertusa.com/index.html
Use the "Desert Directory" to guide you to the "Animal & Wildlife" or "Plants & Wildflowers" section of this Desert USA site to learn more about animal and plant adaptations in the desert.

http://yahooligans.yahoo.com/content/animals/
Want to know more about a particular animal and its adaptive characteristics? Then check out this Ranger Rick site that provides simple information on many mammals, fish, birds, insects, amphibians, and reptiles.

Books

Charles Darwin: And the Evolution Revolution by Rebecca Stefoff. This biography of Charles Darwin explores the scientist, his theories, his time, and his impact. Oxford University Press, 1998. [RL 7 IL 6–12] (5481601 PB)

Evolution by Linda Gamlin. An Eyewitness Book on evolution. Dorling Kindersley, 1993. [RL 8.3 IL 3–8] (5868706 HB)

Books (continued)

Evolution by Alvin and Virginia Silverstein and Laura Silverstein Nunn. Explains a fundamental concept of science, gives some background, and discusses current applications and developments. Millbrook Press, 1998. [RK 5 IL 5–8] (3112006 HB)

Extinct! Creatures of the Past by Mary Batten. Describes giant bugs, birds, and mammals that lived long ago and became extinct during the last ice age, discusses the extinction of more recent animals, and examines the effort to protect endangered species. Golden Books, 2000. [RL 2.8 IL 2–4] (3267001 PB 3267002 CC)

Titles in the Animal Series:

Animal Senses by Michel Barre. Gareth Stevens, 1998. [RL 5.2 IL 3–7] (5890806 HB)

Animals and the Quest for Food by Michel Barre. Gareth Stevens, 1998. [RL 5 IL 3–7] (5890906 HB)

How Animals Move by Michel Barre. Gareth Stevens, 1998. [RL 5.9 IL 3–7] (5892006 HB)

How Animals Protect Themselves by Michel Barre. Gareth Stevens, 1998. [RL 5.8 IL 3–7] (5892106 HB)

•RL = Reading Level
•IL = Interest Level
Perfection Learning's catalog numbers are included for your ordering convenience. PB indicates paperback. CC indicates Cover Craft. HB indicates hardback.

Glossary

adapt (uh DAPT) to change in order to survive in an environment

adaptation (ad ap TAY shuhn) characteristic that helps an organism survive in its environment

algae (AL jay) organisms that live in water and make their own food through photosynthesis (see separate entry for *photosynthesis*)

alpine (AL peyen) relating to a cold upper mountain environment

bacteria (bak TEER ee uh) single-celled organisms

camouflage (KAM uh flahzh) adaptation that animals use to blend into their environment; protective coloring

copepod (KOH puh pahd) tiny freshwater or ocean animal with a hard shell and jointed legs that lives among plankton (see separate entry for *plankton*)

environment (en VEYE er muhnt) set of conditions found in a certain area; surroundings

evolution (ev uh LOO shuhn) process by which species change over time

extinct (ex STINKT) having no members of a species alive

habitat (HAB i tat) place where an organism lives

hibernation (heye ber NAY shuhn) sleeplike state that animals go into during the winter

immune (im YOUN) not affected by; protected from

mammal	(MAM uhl) warm-blooded animal that feeds its young with milk from the mother
mate	(mayt) to come together to reproduce (see separate entry for *reproduce*)
migration	(meye GRAY shuhn) movement from one area to another and back again to adapt to changing environments
mimicry	(MIM uh kree) resembling another animal
offspring	(AWF spring) new organism; child
organism	(OR guh niz uhm) living thing
photosynthesis	(foh toh SIN thuh sis) process by which plants use energy from the Sun to make food
phytoplankton	(feye toh PLANK tuhn) type of plankton that uses the Sun to produce food (see separate entry for *plankton*)
plankton	(PLANK tuhn) tiny organism that floats in a body of water
pollen	(PAH lin) powdery substance produced by plants that contains one of the materials necessary for reproduction (see separate entry for *reproduce*)
predator	(PRED uh ter) animal that hunts other animals for food
prey	(pray) animal that is hunted by other animals for food
reproduce	(REE pruh doos) to make more organisms of the same species
rodent	(ROH dent) gnawing animal
species	(SPEE shees) group of living things that resemble one another and can reproduce to form more members of the group
variation	(vair ee AY shuhn) difference in an organism within a species

Index